Where Babies Come From

Rosemary Stones • Illustrated by Nick Sharratt

Do you know where babies come from? Sometimes children are told that babies are found under a gooseberry bush or that the stork brings them or that there is a special baby shop where you can buy them. These are just funny stories which are not true. This book tells you where babies *really* come from.

BABY SHOP

Have you wondered why girls' bodies and boys' bodies are different? It's so that when girls and boys grow up and become women and men, they can be mums and dads if they want to. The different bits are called the sex parts or genitals.

Between their legs boys have a small bag called the scrotum, which contains their testicles. In front of the testicles is a tube called the penis. When a boy needs to pee, the pee comes out through his penis. Penises don't all look exactly the same.

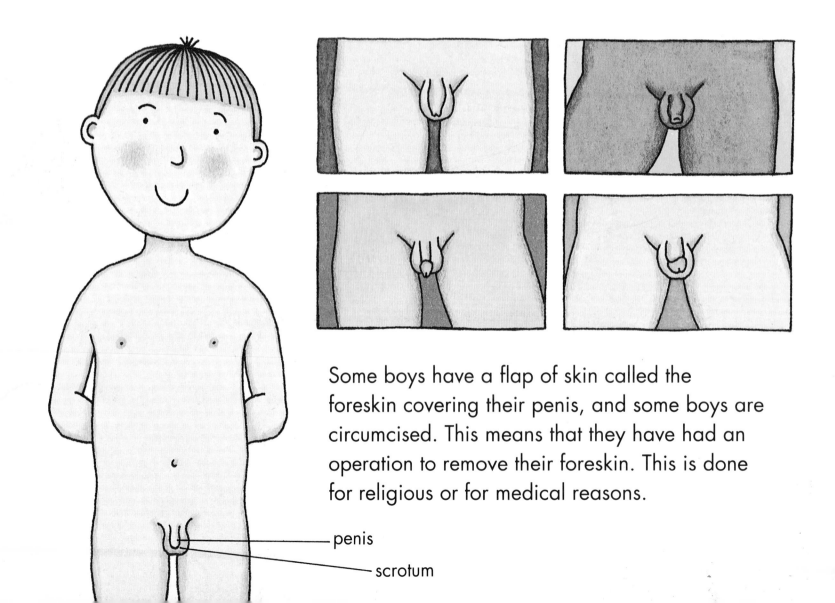

Some boys have a flap of skin called the foreskin covering their penis, and some boys are circumcised. This means that they have had an operation to remove their foreskin. This is done for religious or for medical reasons.

penis

scrotum

Between their legs girls have two lips called the vulva. The vulva covers a special opening called the vagina and a small tip of flesh called the clitoris. It also covers a very tiny opening called the urethra. When a girl needs to pee, the pee comes out through this opening. Vulvas don't all look exactly the same.

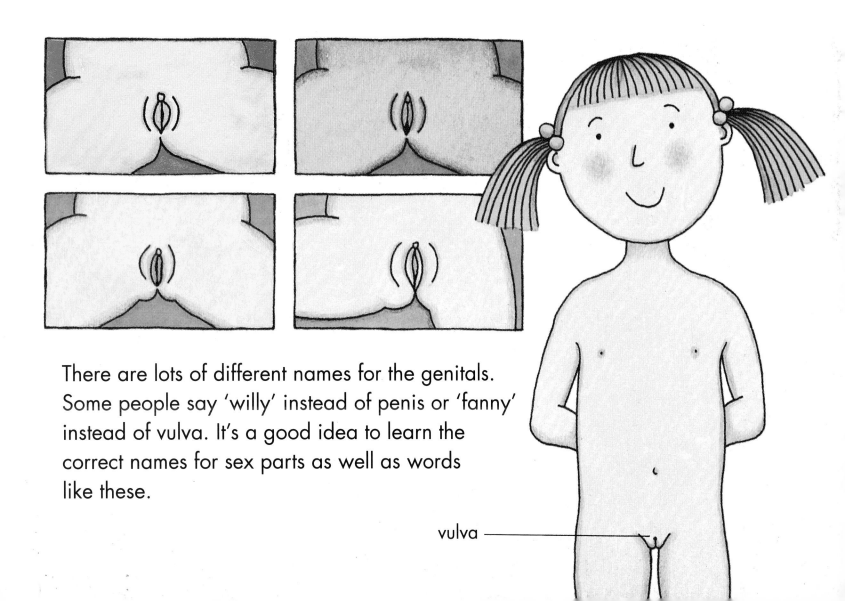

There are lots of different names for the genitals. Some people say 'willy' instead of penis or 'fanny' instead of vulva. It's a good idea to learn the correct names for sex parts as well as words like these.

vulva

When girls reach the age of about nine or ten their bodies begin to change and develop into grown-up bodies. Boys' bodies usually start changing a bit later when they are eleven or twelve. These changes take place gradually over a number of years.

A girl's breasts begin to appear and hair grows round her vulva and in her armpits.

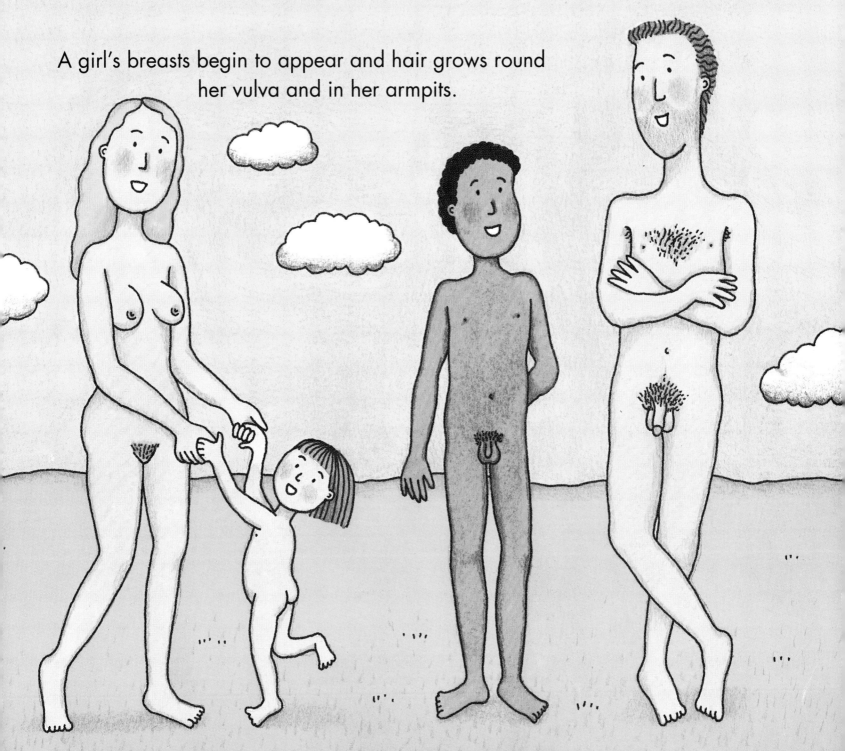

A boy's penis and testicles get larger and hair grows around them and in his armpits. Hair also starts to grow on his face.

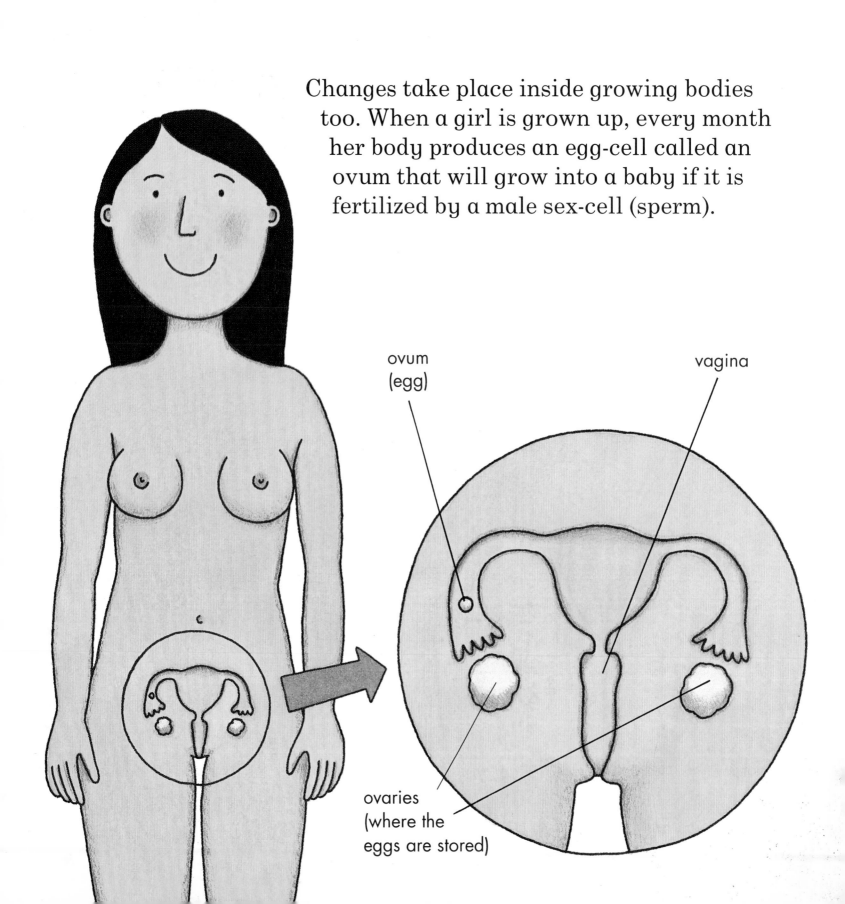

Changes take place inside growing bodies too. When a girl is grown up, every month her body produces an egg-cell called an ovum that will grow into a baby if it is fertilized by a male sex-cell (sperm).

ovum (egg)

vagina

ovaries (where the eggs are stored)

When a boy is grown up, his testicles begin to produce male sex-cells called sperm. Sperm is needed to fertilize a woman's ovum if a baby is wanted.

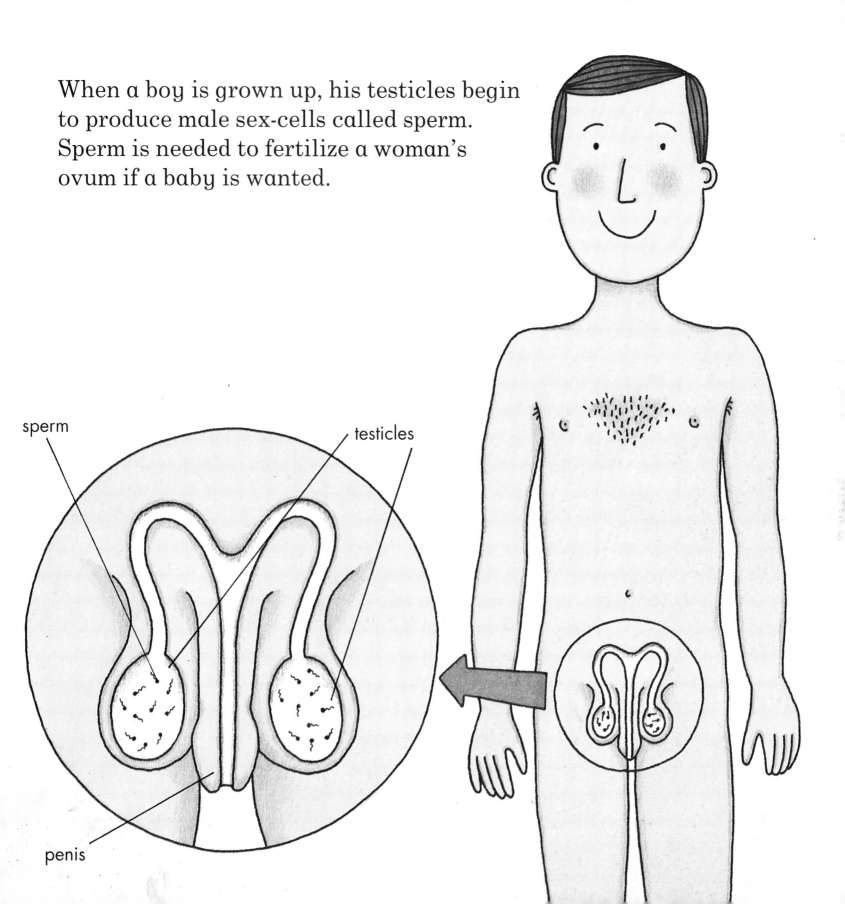

sperm

testicles

penis

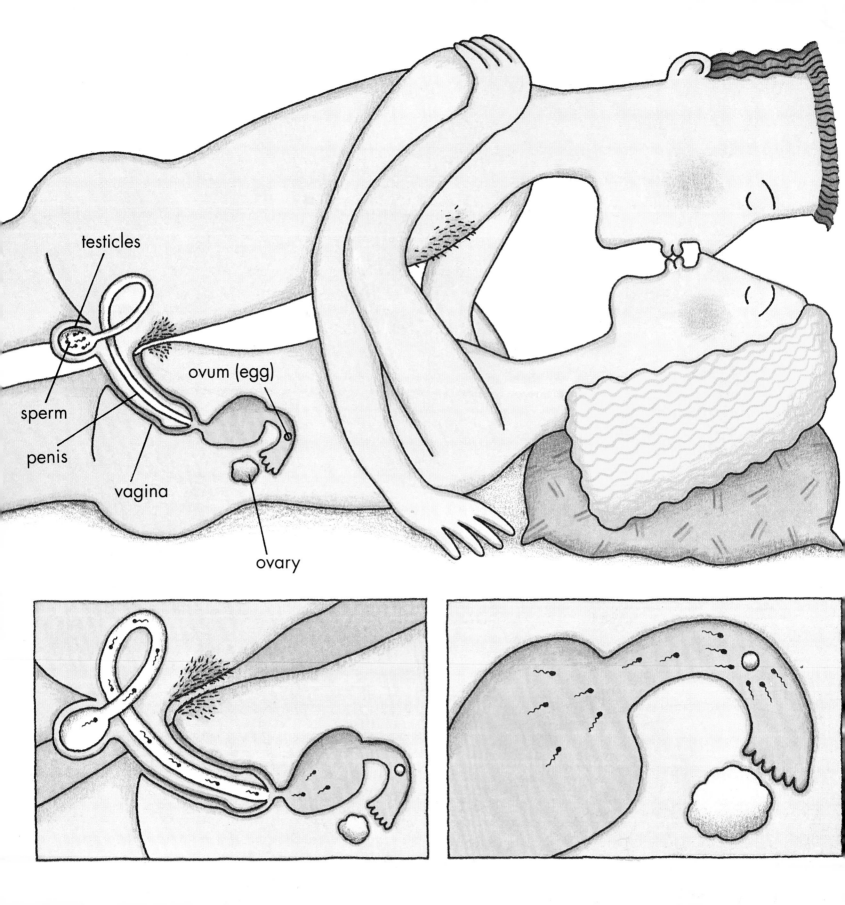

testicles

sperm

penis

vagina

ovum (egg)

ovary

When a woman and a man decide that they would like to have a baby, they have sexual intercourse. This is also called making love and it feels very exciting and loving and good. They kiss and touch each other until the woman's vagina becomes wet and slippery and the man's penis becomes stiff. They then hold each other very close so that the man's penis slides inside the woman's vagina. Eventually sperm travels from the man's testicles along his penis and out into the woman's vagina. One of the sperm cells fertilizes the woman's ovum and this is the beginning of a baby.

A sperm joining with the egg to fertilize it.

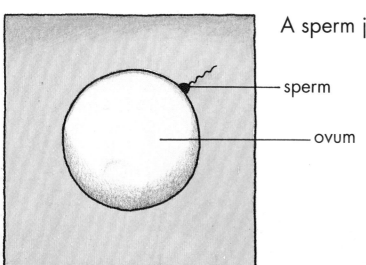

sperm

ovum

If grown-ups want to make love and not have a baby they use birth control (contraception). This is a way of stopping the woman's ovum from being fertilized.

The tiny fertilized ovum grows inside the mother's tummy in a special bag called the womb. At first the baby is no bigger than a tadpole, then its head, arms and legs begin to grow. The growing baby is fed from the mother's body through a cord which grows from the middle of its tummy (the umbilical cord).

If you look at your tummy you will see your tummy button. This is where your umbilical cord grew when you were inside your mother's womb. When you were born the cord was cut.

Will the baby be a boy or a girl? Usually we don't know until the baby is born. Sometimes doctors do tests to find out if a baby is growing properly and they find out the sex of the baby before it is born.

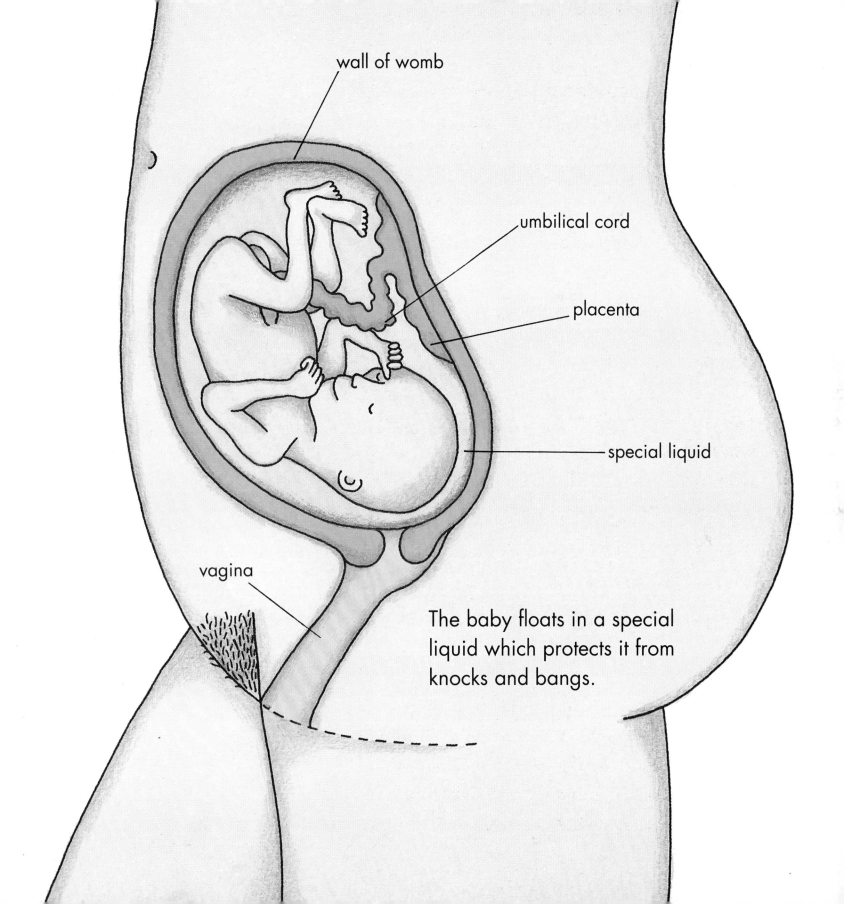

wall of womb

umbilical cord

placenta

special liquid

vagina

The baby floats in a special liquid which protects it from knocks and bangs.

After about four months, the mother can feel the growing baby moving inside her. At first it's a fluttering feeling. When the baby gets really big, the mother can see her tummy move when it kicks. If your mum or aunt is pregnant (expecting a baby), ask if you can put your hand on her tummy when the baby starts kicking.

As the baby grows, the mother's tummy gets very big. She finds it difficult to bend down to put her shoes on and it is not good for her to carry heavy things. If your mum is pregnant she will be pleased if you help by fetching and carrying things for her.

After about nine months the baby is ready to be born. The muscles of the mother's womb contract and push the baby down through her vagina and out through the opening between her legs. This opening is very elastic and it stretches to let the baby out.

Some babies are born in hospital and some babies are born at home. The mother is helped during the birth by a midwife or a doctor. Quite often dads like to be there too, to see their baby being born. Some say it was the most exciting moment of their life!

A midwife is specially trained to help mothers and deliver babies.

Most babies are born head first, but some are born bottom first. Sometimes a baby is in such an awkward position that doctors have to do an operation to take out the baby safely through a cut in the mother's tummy.

After the baby is born it is weighed and measured. The baby's growth is checked for several months to make sure that it is doing well.

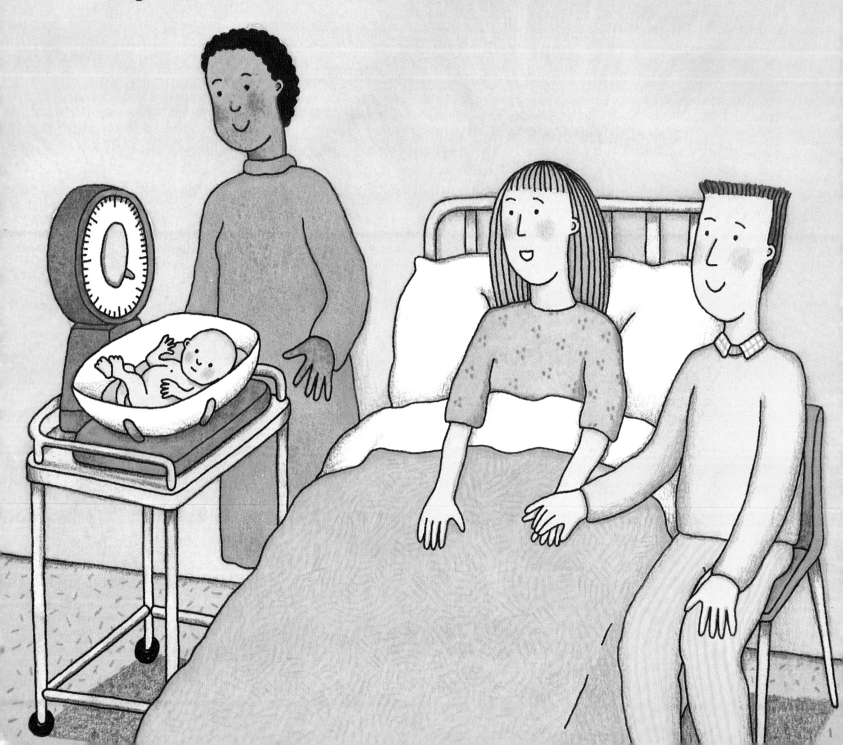

After the birth the mother and baby feel very tired. It's hard work having a baby and it's hard work being born.

If a baby arrives too early it may have to stay in an incubator and have special care until it has grown big and strong.

New babies love being talked to and they love cuddles, but you must always hold a baby carefully, supporting its head.

New babies have a surprisingly strong grip. Put your finger gently into a baby's fist and you will find that the baby grips it tightly.

In some parts of the country it is the custom to put a silver coin into a new baby's fist. If the baby grips it, it will grow up to be rich.

Many mothers like to breastfeed their baby. The baby sucks the mother's breasts to draw out the milk that her body has made.

Some babies are fed from a bottle with a teat. Dads and older brothers and sisters can bottle-feed babies too.

New babies often wake up in the middle of the night and cry to be fed. Mums and dads can get very tired and bad-tempered. But it won't always be like this.

Soon the new baby will learn to smile, to sit up, to wave, to make funny noises and eventually to walk and talk.

One day the new baby will be as big as you.